S0-DTB-535

by Patricia Lauber
Illustrated by Martha Weston

THOMAS Y. CROWELL NEW YORK

What Big Teeth You Have!

10 9 8 7 6 5 4 3 2 1
First Edition

Library of Congress Cataloging-in-Publication Data
Lauber, Patricia.
What big teeth you have!

Summary: Explains how teeth are clues to what animals eat and how they get their food.
1. Teeth—Juvenile literature. 2. Animals—Food—Juvenile literature. [1. Teeth. 2. Animals—Food habits] I. Title.
QL858.L38 1986 596′.0132 85-47902
ISBN 0-690-04506-9
ISBN 0-690-04507-7 (lib.bdg.)

WHAT BIG TEETH YOU HAVE!

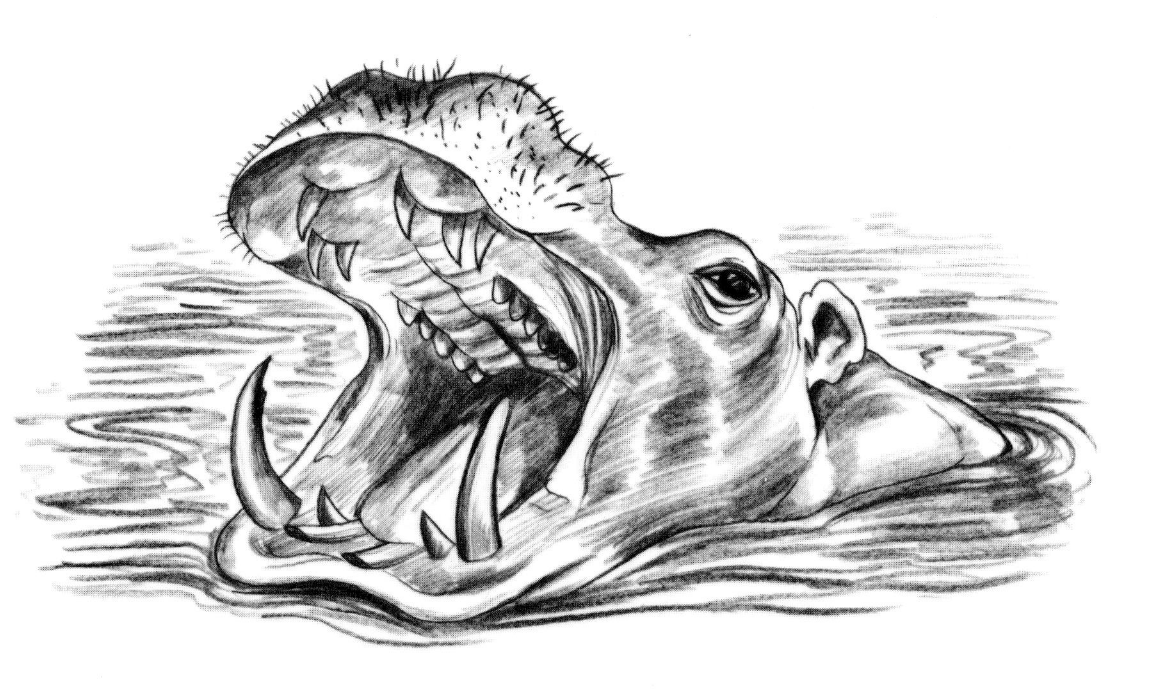

What Big Teeth You Have!

Look at these animals and their teeth. Your first thought may be that the teeth are very big. That is so, but something else is more important. Look closely and you will see what it is. No two sets of these teeth are alike. The beaver, for example, has very big front teeth. The giraffe has no upper front teeth at all.

If you know how, you can tell a lot about an animal by looking at its mouth and teeth. They are clues to what it eats and how it gets its food. This is true of mammals, such as beavers, hippos, horses, lions, and giraffes. It is also true of reptiles and fishes.

Nearly all mammals have teeth. Most kinds use their teeth to catch or gather food. Only a few have hands or paws that can do this work. Teeth are the chief tools that mammals use to get food and start it toward the stomach. Teeth may also be used for fighting, grooming, or making a home.

Like tools, teeth have many shapes. Each shape is suited to doing a certain kind of job. That is why teeth are clues to what an animal eats and how it gets its food.

There are four main types of mammal teeth. You can see them all by looking in a mirror and opening your mouth.

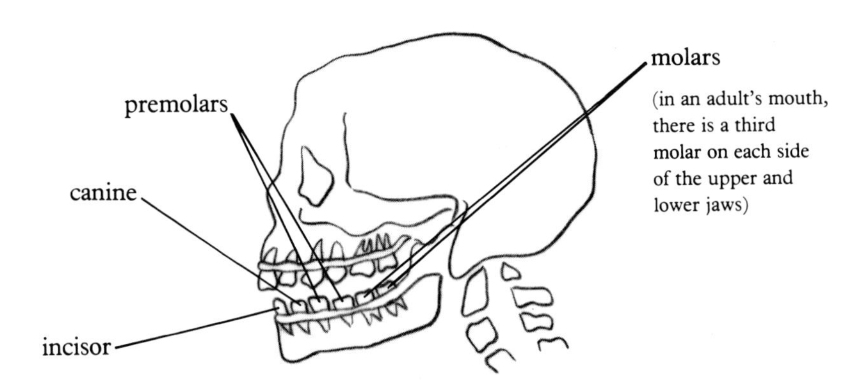

The names come from Latin words that tell something about the teeth.

English name	*Latin word means*	*What the teeth do*
incisor	to cut	Incisors are cutting teeth. You use yours to cut off mouthfuls of food.
canine	dog	The dog family has big sharp canine teeth, used for stabbing and tearing. Our small canines are used for cutting and holding.
molar	millstone	Millstones were used for grinding and crushing grain. Molars grind or crush food. (**Premolars**, which come before molars in your mouth, both cut and grind.)

Wolves, Horses, and Chimps

Like you, a wolf has the four main types of mammal teeth. But a wolf's teeth and jaws are bigger and stronger than yours. The teeth are very sharp and the canines are long. These are the teeth of a hunter.

Wolves hunt in packs when they are after a big animal, such as a moose. They chase or stalk their prey and attack with a rush. Their long canines stab deep. Their incisors act as clamps. The wolves hang on and drag the prey down. If there are no big animals, a wolf hunts small ones.

You can see how wolves eat by watching a dog with a meaty bone. Wolves and dogs are relatives. They have the same sort of teeth and they eat in the same way.

A dog uses its incisors to nibble or tear off bits of meat. The premolars work like heavy scissors, or shears. A dog uses them to slice through tough parts. It also uses them for gnawing. It uses its molars to crack and crush bones.

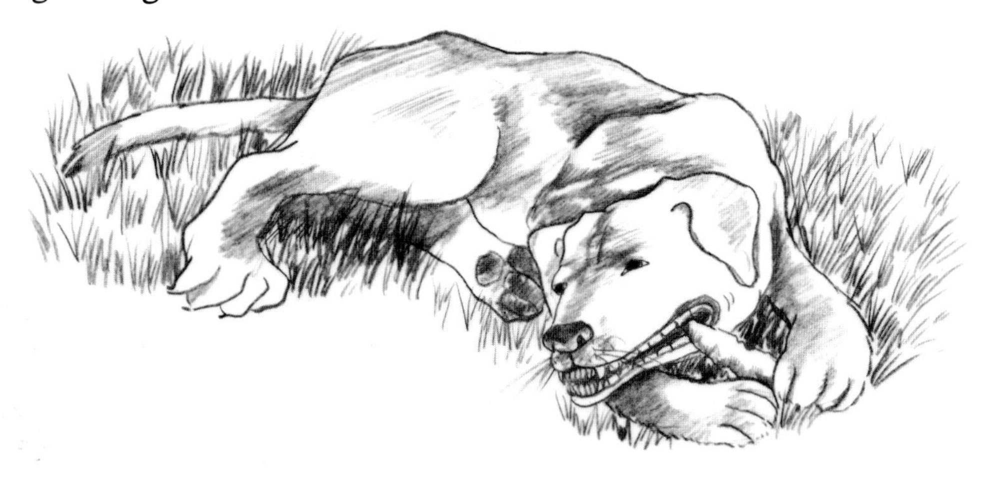

The cat family is also made up of hunters and meat eaters. Their teeth are much like those of wolves, but even sharper. Most of the cats do not chase their prey. They hide or creep up on it—and pounce. Lions, for example, cannot run fast or far. Zebras can. To catch a zebra, a lion must take it by surprise.

Horses are grazing animals. They eat grass and other plants.

What kind of teeth does it take to eat grass? The answer may surprise you. It takes big strong teeth. Grass is a gritty, harsh food. It wears down teeth quickly. If a horse had small teeth like ours, they would wear out in a year or so.

If you look a horse in the mouth, you will see its incisors. They are big and slightly curved. They look like chisels. These teeth are good for cutting through tough, wiry grass.

A horse uses both its premolars and its molars for grinding. Together they are called the cheek teeth. Each cheek tooth is like a column with a broad top. The top has ridges that help to grind grass to a pulp.

The grinding wears down even big, strong teeth. But the cheek teeth keep pushing out from the gums until, in time, they wear right down to their roots. By then the horse has usually lived as long as it is going to.

Most of the time giraffes reach high in the air for food. They eat leaves and tender shoots from the tops of trees.

A giraffe has no upper incisors or canines. In place of these teeth, there is a hard, thick pad. A giraffe eats by wrapping its long black tongue around leaves and pulling them into its mouth. It cuts off a mouthful of food with its lower incisors and canines, which are flat, not pointed.

Like a cow, a giraffe chews its food lightly and swallows it quickly. The food goes into the stomach, which has four parts. The food is broken down and softened in the first two parts. It becomes a pulp. Later the giraffe brings up the pulp, which is called a cud. This time the giraffe chews its food thoroughly with its cheek teeth. Unlike a cow, a giraffe chews its cud standing up. Upright a giraffe is safe from lions. It is not safe when it lies down or spreads its front legs to feed from the ground. A giraffe cannot get up quickly.

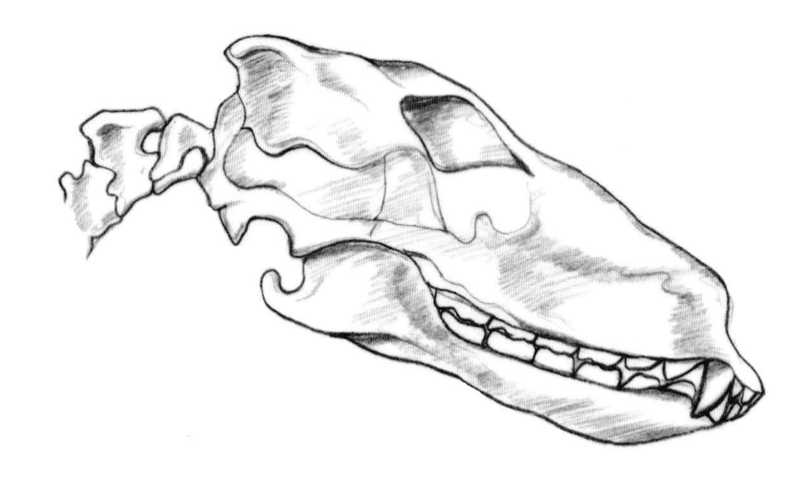

What does the owner of these teeth eat? The big pointed canines tell you that it hunts animals and eats meat. But look at the molars. They are much flatter than a wolf's. They are suited to grinding up plant food. The owner eats both plant food and animal food. It is a bear. This one is a black bear, the most common kind in North America.

A black bear seldom catches a big animal. It is too slow and clumsy. It eats dead animals that it finds. It digs out the burrows of small animals and eats chipmunks, mice, woodchucks, ground squirrels. Most of the time the black bear eats plant foods, filling up on what is in season. In spring it eats grasses and clovers. It eats the tender inner bark of trees. It digs for roots and bulbs. Later in the year it eats all kinds of berries, fruits, and nuts. It may climb a tree to shake down apples or nuts.

Like bears, raccoons have teeth that are good for nibbling, stabbing, cutting, and grinding. A raccoon can eat fish and frogs and small land animals. It can also eat nuts, fruits, seeds, corn. In getting its food and eating, a raccoon often uses its paws. It is almost as skilled in this way as monkeys and apes are.

Chimpanzees also have the teeth of mammals that eat both plant and animal food. Even so, scientists used to think that wild chimps ate only plant food. Then a few scientists spent years watching wild chimps in Africa. They learned that chimpanzees do hunt and eat other animals. They chase or stalk young bushpigs, antelopes, and baboons. They eat birds, lizards, and insects. A chimpanzee often shares its kill with other chimpanzees.

Chimps eat many kinds of plant foods—leaves, fruits, flower buds, grasses, nuts, seeds, mushrooms. They gather their food with their hands and carry it to their mouths. There it is nibbled or cracked, crushed, and chewed.

The world of mammals is a big one. There are many kinds of mammals. Some have teeth like those of wolves or horses or bears. Others do not. They may have teeth with very special shapes. They may even have no teeth at all.

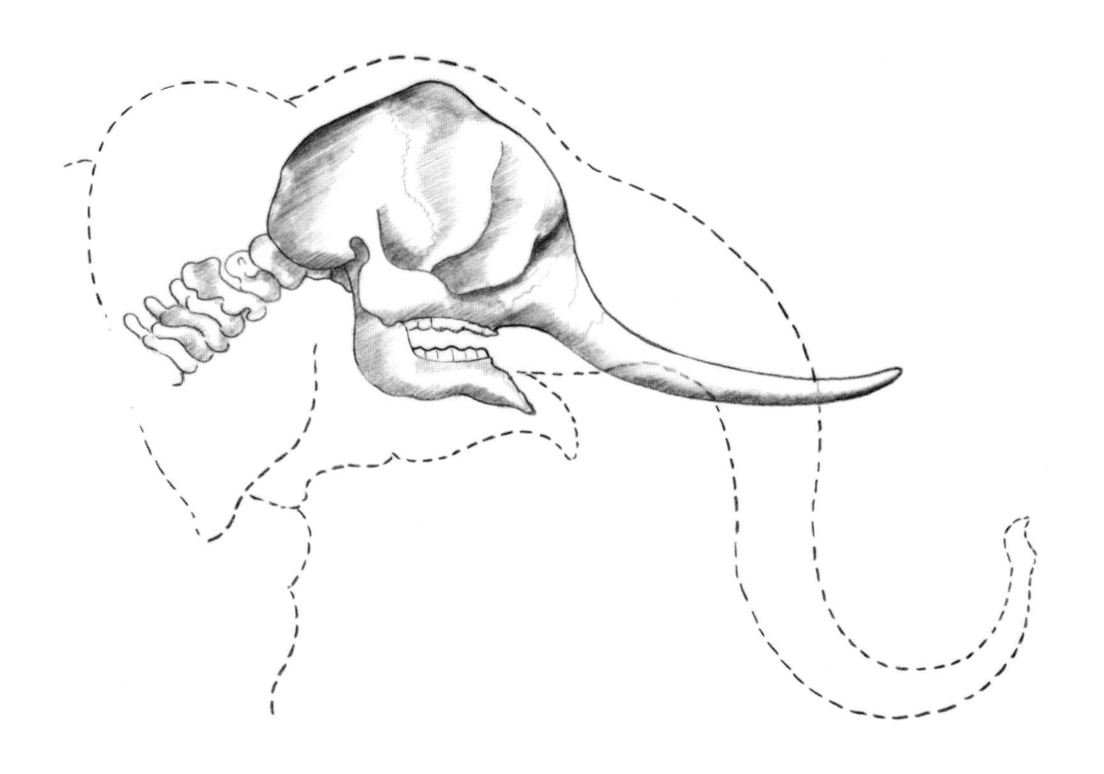

Elephants, Hippos, and Walruses

An elephant is a big animal with big teeth. Its tusks are really teeth. They are its two upper incisors, and they grow all through an elephant's life. An old African elephant may have tusks that are ten feet long and weigh 125 pounds each.

The tusks are deadly weapons. With them, an elephant can even kill a lion that is attacking its young. Tusks are also used in getting food—in rooting plants from the ground or breaking branches from trees.

Elephants need a lot of food, and they spend much of their time eating. Usually an elephant uses its trunk to gather leaves, fruits, vines, grasses, and other plant food. The trunk carries food to the mouth. There it is thoroughly chewed by huge cheek teeth. The rear molars in an Asiatic elephant may be a foot long.

A young adult elephant has 24 cheek teeth, six in each half of each jaw. But it uses only the front four to grind up

its food. In time, these teeth wear down. When a tooth can no longer be used, it is pushed out by the tooth behind it. That tooth moves forward, taking the place of the lost tooth. Then the elephant has five cheek teeth in that part of its jaw.

The same thing happens again and again until the sixth tooth is up front and the only one left. By that time the elephant is very old.

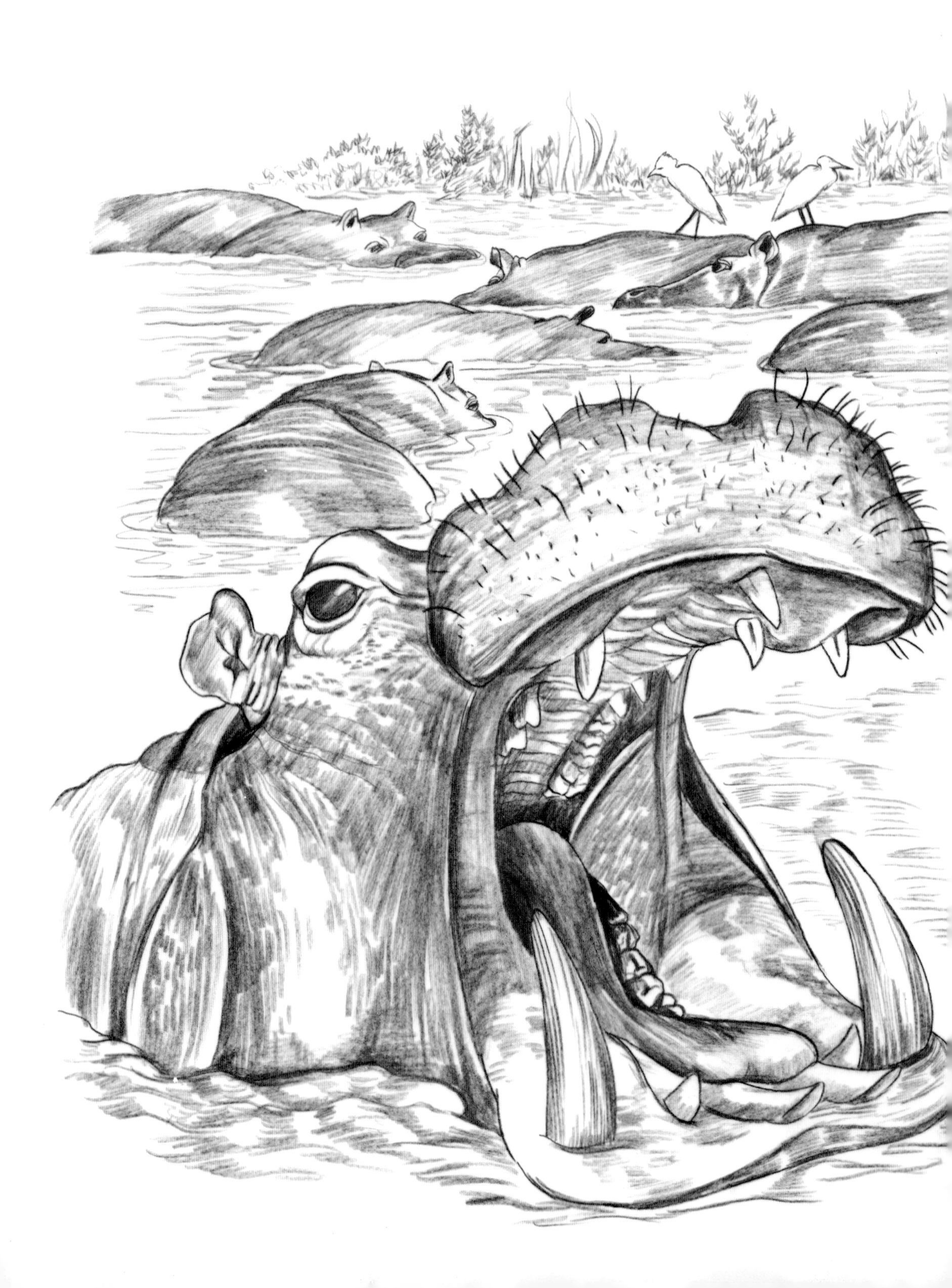

You are looking at one of the biggest mouths in the world. It belongs to a hippopotamus. The teeth look dangerous—and they are. The big lower canines are like tusks and they serve as weapons. A hippo will attack anything that seems threatening. It can kill a smaller animal with one bite. But it does not kill for food. As you can see by the grinding teeth, hippos eat grass and other plants. They pull and cut their food with their lips and lower incisors.

Perhaps you are wondering how an animal with such big canines can close its mouth. The answer is that there are two pockets in the upper jaw. When a hippo closes its mouth, the lower canines fit into these pockets.

Here is a mammal with no teeth. It is the giant anteater of Central and South America. Its chief foods are ants and termites, which it gathers on its long, sticky tongue. It tears open an ant nest with its claws and pushes its tube-shaped head into the opening. It sticks out its long, wormlike tongue. The scurrying ants stick to the tongue, and the anteater pulls its tongue into its mouth. It may swallow thousands of ants at a single meal.

Walruses live along the shores of northern seas. They spend part of their time in the water and part of their time out of the water. Sometimes they haul themselves out onto floating ice. To do this, a walrus uses its tusks. It jabs them into the ice and heaves itself out of the water.

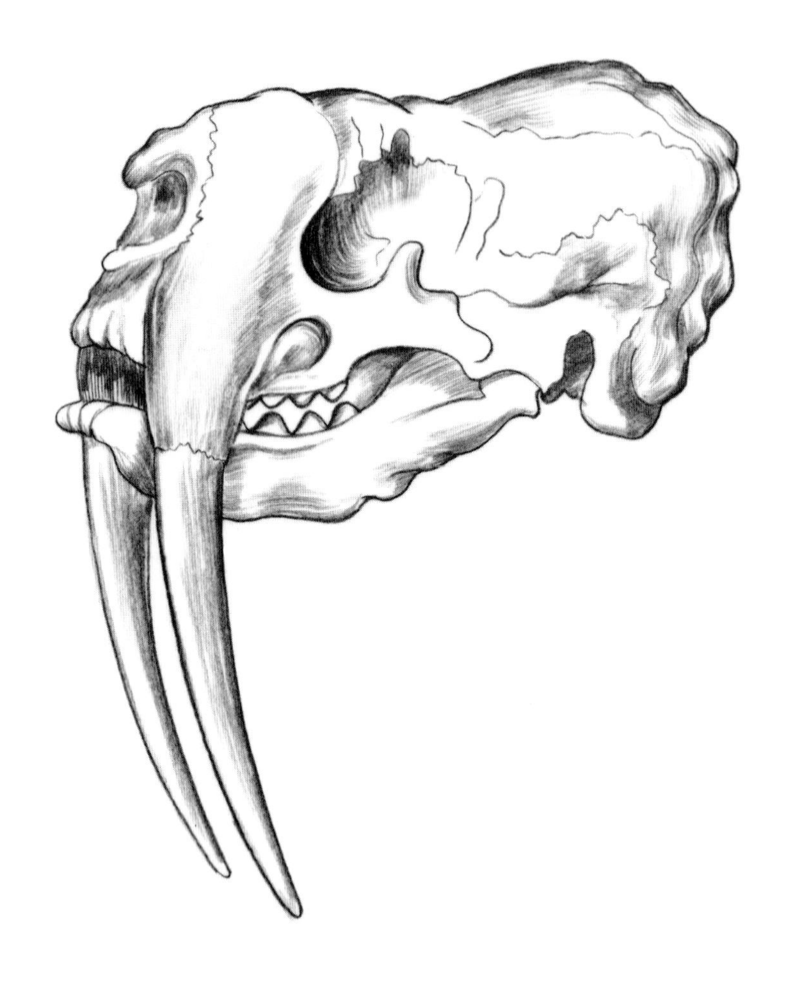

The tusks are teeth. They are the two upper canines, and they grow all through a walrus's life. An old walrus may have tusks that are 40 inches long. The walrus is a big, strong animal, and its tusks are dangerous. Most other mammals keep away from the walrus. Even polar bears must be careful.

A walrus finds its food on the sea bottom. Clams are its main food. How does it eat? So far no one knows.

Perhaps a walrus crushes shells with its molars. These are shaped like cones with blunt ends. That is a good shape for crushing shells. But there is a problem with this idea. Not even tiny pieces of shell have ever been found in a walrus's stomach.

Perhaps the walrus feeds in another way. It might find food and then suck off the soft parts with its strong lips. That would explain why there are no pieces of shell in the stomach. But why, then, does a walrus have big molars that are shaped for crushing shells? Some future scientist may solve the mystery.

The teeth of a walrus are puzzling, but most mammal teeth are not. They tell what mammals eat and how they get their food. And there are still other ways that scientists use them. Teeth are a big help when scientists are sorting mammals into groups of relatives. They show, for example, that the beaver has many relatives you might never think of.

Beavers, Bats, and Whales

A beaver has four long incisors, two above and two below. It has no canines. There is a big space between the incisors and the cheek teeth.

Beavers use their incisors to cut off plant food—grass, roots, herbs, tree bark. A beaver may cut a branch off a fallen tree. It nibbles along the branch, like a person eating corn on the cob. It eats the buds, twigs, and tender green bark. All food is ground up by the beaver's cheek teeth.

Many beavers use their teeth to make homes for themselves. They build dams to make ponds. Then they build lodges in the ponds. In its lodge, a beaver family is safe from enemies.

To build, a family must first fell trees.

Choosing a tree, a beaver stands on its hind legs and braces itself with its tail. It grips the tree with its upper incisors and cuts out a chip with its lower incisors. Two or three inches below the first chip, it cuts out a second one. Then it tears out the middle piece of wood with a strong jerk of its head and teeth.

Working in this way, the beaver circles the tree. It cuts deeper and deeper until the tree falls. After a rest and something to eat, the beaver gets back to work. It starts cutting up the tree and dragging pieces to the dam or lodge it is building.

You might think a beaver's incisors would soon wear out. But this does not happen. They keep growing all through the beaver's life.

The incisors never become dull, either. They are self-sharpening. The front of each tooth is harder than the back. The back wears away faster, leaving a sharp edge.

Beavers have many relatives. Among them are squirrels, chipmunks, rats, mice, gerbils, hamsters, porcupines, and guinea pigs.

What makes them relatives? Their teeth. All of these animals have four big self-sharpening incisors that grow throughout their lives. That is why scientists have put them together in one large group. The animals are all rodents, or gnawing mammals. Animal teeth are important clues for the scientists who work at sorting out animals, or classifying them.

It's easy to sort out bats from other mammals. Bats are the only ones with wings. They are the only mammals that can truly fly. All told, there are between nine hundred and a thousand kinds of bats. They are not easy to sort out into smaller groups. But one way of doing it is by their teeth and what they eat.

The little brown bat of North America has small, sharp teeth. Its cheek teeth are like steak knives. These are the teeth of an insect-eating bat. Insects do not have bones. Instead, their bodies are covered with a hard material. It takes sharp teeth to pierce and chop up insect bodies. By far the largest number of bats are insect eaters.

This bat is called a flying fox because of its fox-shaped head. The flying fox is one of the fruit-eating bats. It crushes its food between its flat molars, swallows the juice, and spits out the seeds and pulp. Some fruits have hard or tough husks. A flying fox uses its long canines to pierce husks and break them open. It can even open a small green coconut.

A baby flying fox has teeth that are hooked on the ends. When the mother flies out to feed, the baby goes, too. It clings to her with its teeth and claws.

There are many kinds of flying foxes in Australia and Southeast Asia. In these warm regions they find fruits all year round.

Still other kinds of fruit-eating bats live in Africa.

All the long-nosed bats live in warm places. A long-nosed bat has the teeth of a fruit eater, but it also eats nectar and pollen. It collects them from flowers with its long, bristly tongue.

This small bat has very large upper incisors. They are as sharp as razor blades. Its cheek teeth are tiny and the bat never uses them. It doesn't need to. Its only food is blood. This is one of the vampire bats of Central and South America.

The vampire flies out at night looking for a quiet, warmblooded animal. It walks lightly and gently over the sleeping victim, searching for a spot where there is little or no hair. It uses its razor-sharp incisors to shave off a thin slice of skin. Blood wells up. The vampire drinks about an ounce of blood, then silently flies away.

Some bats eat frogs, lizards, birds, and small mammals. They even eat other bats—they are cannibals. This one lives in Australia. It is a large bat and so pale that it is called the ghost bat.

Look at the teeth. The ghost bat's upper canines are big, and they point slightly forward. These are slashing teeth. The ghost bat catches its prey with these teeth. Then it flies to its favorite eating place. It hangs upside down by its feet and eats its meal.

Dolphins, porpoises, and whales are relatives. Scientists divide them into two main groups: those with teeth and those without teeth.

Each animal in the toothed group has teeth that are all the same shape.

The bottle-nosed dolphin has 80 to 104 teeth. They are all shaped like cones, and they point forward. These teeth are good for catching slippery fish. They are not good for chewing, and a dolphin doesn't chew. It swallows its catch whole.

Dolphins and porpoises are close relatives and look-alikes. The quickest way to tell them apart is by the shape of their teeth. A porpoise has spade-shaped teeth.

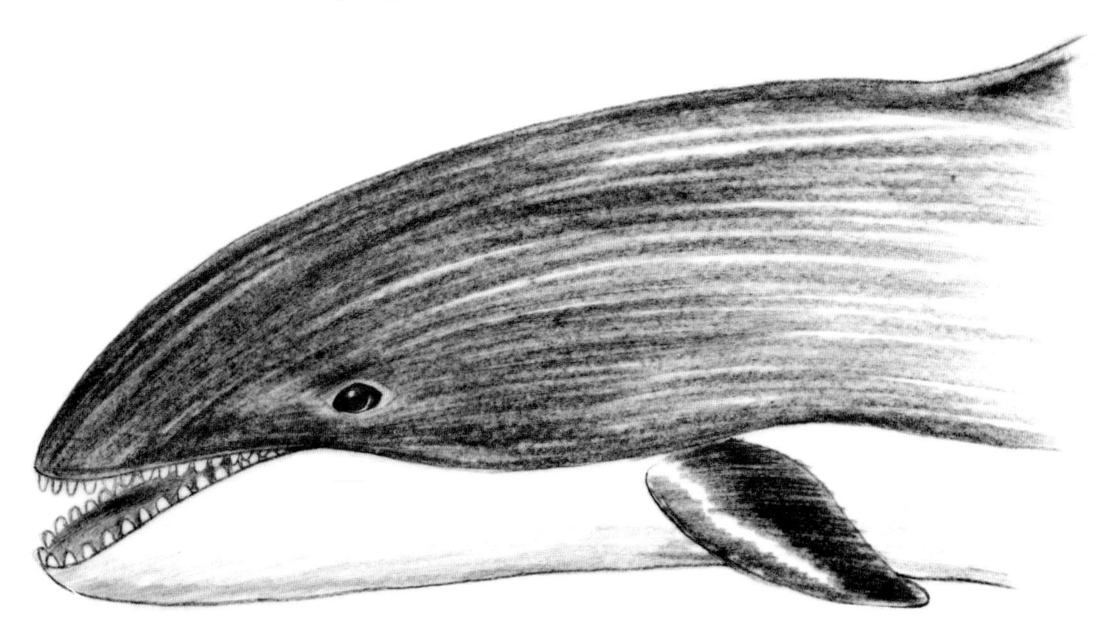

The narwhal is a whale that lives in Arctic seas. It has only two teeth. Both are in the upper jaw. In young narwhals the two teeth are alike. But in adult males one tooth grows into a long tusk that may be eight feet long. It has the shape of a spiral. No other animal has a spiraled tooth.

How is this tooth used? Scientists who study whales are not sure. It doesn't seem to be used for breaking ice. It isn't used for spearing food—cod, shrimp, and squid. Most likely, the scientists think, the males use their tusks as swords to fight for females at mating time. Many males have scarred faces, and many have broken tusks.

Of all the giant whales, only one kind has teeth. This is the sperm whale. All its teeth are in the lower jaw. They are all alike and each is about the size of a man's fist.

When the sperm whale closes its mouth, the teeth fit into pockets in the upper jaw. This whale uses its teeth in catching squid, its main food. It swallows the squid whole.

All the other giant whales belong to the second main group—they have no teeth. Instead, each has what looks like a huge mustache in its mouth. The mustache is made of a material called baleen.

The blue whale feeds on tiny sea creatures known as krill. It swims through a patch of krill-filled water with its mouth open. The mouth fills with water and krill. The whale then closes its mouth and raises its tongue. The water squirts out through the baleen, which acts as a strainer. The krill is left in the mouth and the whale swallows it.

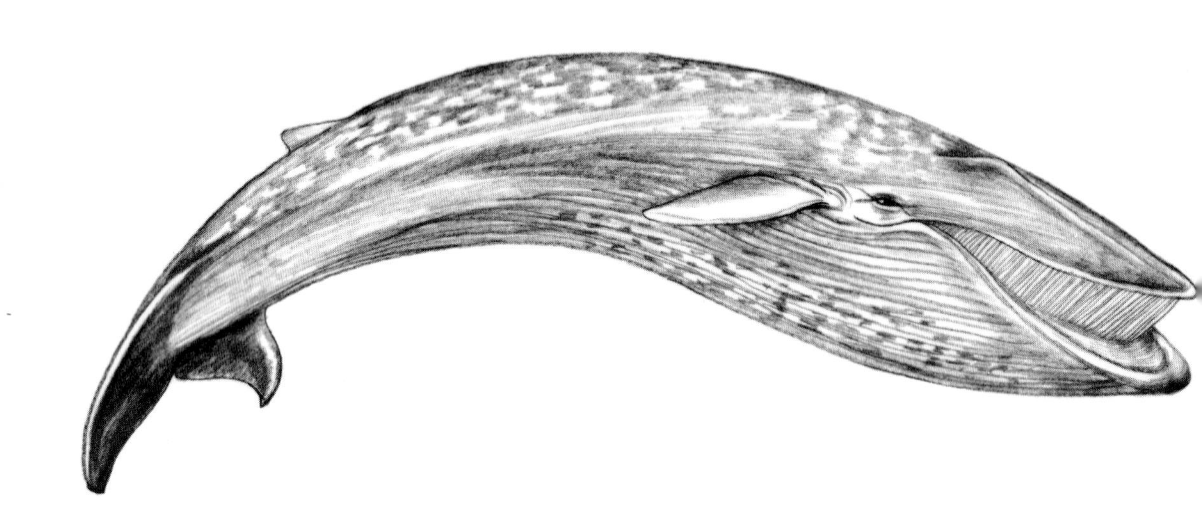

Gators, Snakes, and Sharks

Mammals are not the only animals with teeth. Most kinds of reptiles and fishes also have teeth.

An alligator has lots of teeth, as many as 40 in each jaw. It never runs short. If a tooth breaks off, a new one grows in and takes its place.

As you can see from the teeth, gators eat other animals. They eat whatever they can catch—fish, snakes, turtles, small mammals, and birds. The pointed front teeth are good for catching prey. The blunt back teeth are good for crushing it.

Food is swallowed whole unless it is too big. In that case, the gator tears it to pieces, then crushes and swallows it. Alligators are messy eaters. A very young alligator often feeds on bits of food that fall out of its mother's mouth. It also eats insects and small fish.

Big male alligators may fight over mates. They use their strong jaws and sharp teeth in battle. One may try to tear a leg off the other. Or they may lock jaws and wrestle. The stronger male breaks the jaw of the weaker male.

After mating, a female makes a nest for her eggs. She does much of her work with her teeth and jaws. She rips out clumps of plants and carries them to the nest she is building.

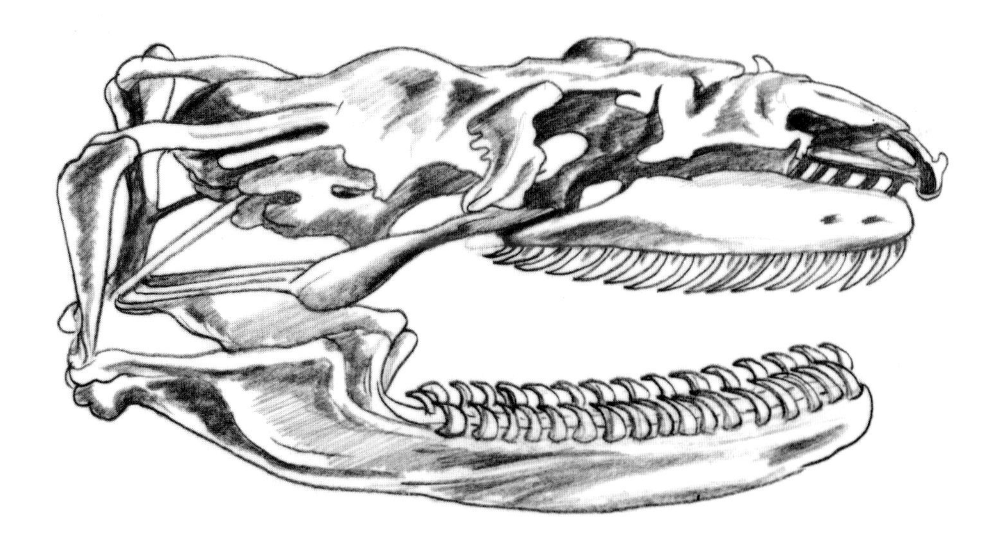

These needle-sharp teeth belong to another hunter. They are the teeth of a snake. A snake's jaws are edged with teeth. Most kinds of snakes also have two more rows of teeth that grow from the roof of the mouth. Snakes use their teeth to catch prey and pull it into their mouths. They always swallow their meals whole. Yet the meal is often bigger than the snake's head!

The yellow rat snake eats eggs as well as rats, mice, and other animals. This one is about to swallow a hen's egg. The job looks impossible, but it isn't. Snakes have jaws that stretch.

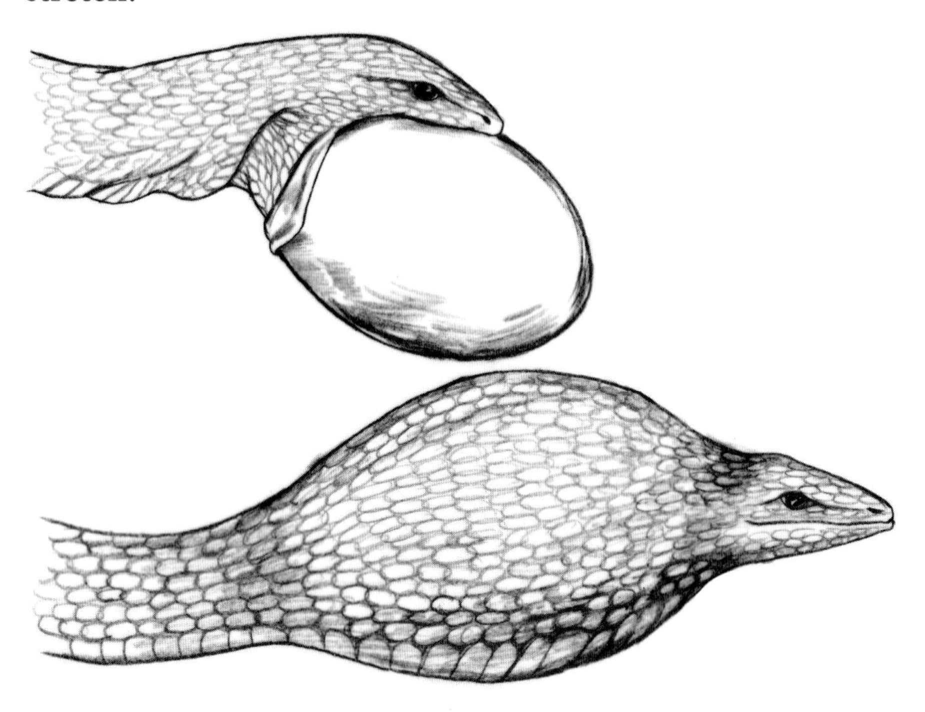

A snake can open its mouth very wide.

Your lower jaw is one piece of bone. A snake's is in two pieces. The pieces can be stretched away from each other.

Also, a snake can drop its lower jaw much farther than you can.

Bones in a snake's upper jaw move, too.

This king snake is swallowing a water snake. Left, right, left, right—the pieces of the upper jawbone move in turn, pushing the teeth forward over the water snake. In this way the water snake is pulled into the king snake's mouth and throat.

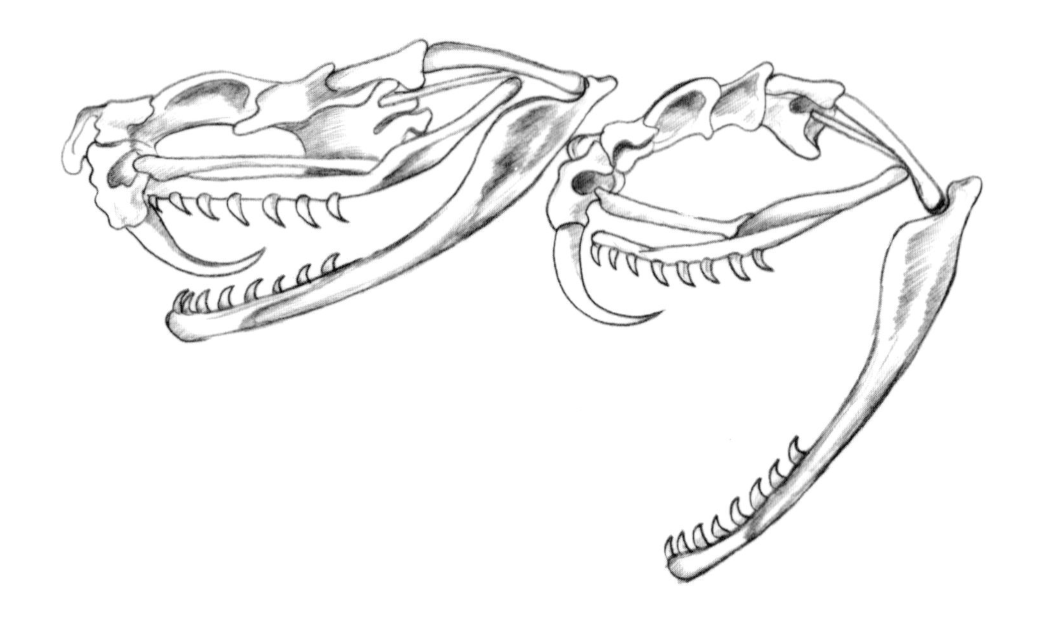

A rattlesnake has two special teeth. These are long fangs in its upper jaw. Each is mounted on a piece of bone that moves. When the fangs are not in use, they are folded back in the mouth. When the rattler opens its mouth to strike, the fangs spring up. The rattler uses its fangs to poison its prey.

The fangs are hollow. Each is linked to a sac of poison in the upper jaw. When the rattler strikes, muscles squeeze the sacs. Poison is forced through the fangs and into the prey.

Sometimes a rattler breaks a fang on a stone or a branch. This does not matter. A rattler has several spare fangs. One of them moves forward and is ready for use.

The many kinds of fishes live in seas, lakes, and streams. They eat many kinds of food, and they have many kinds of teeth. These teeth may grow around the jaw. They may also grow on the tongue, on the roof of the mouth, or in the throat. It all depends on the kind of fish.

A parrotfish has teeth that look as if they had melted together. They form a kind of beak at the front of the jaw. The parrotfish uses these teeth to bite off pieces of plants. Its throat is paved with teeth that grind up the food.

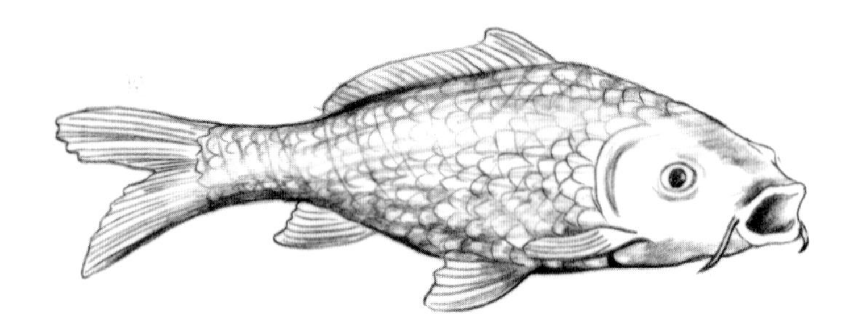

The carp is another plant eater, although it also feeds on shrimp, worms, insects, and smaller fishes. It has no teeth in its jaws, but it does have strong teeth in its throat that grind up food.

The pike is a fish eater, with teeth like spikes. When a pike gulps in its prey, the teeth bend backward. Then they stand up, forming a cage. The prey cannot swim out of the pike's mouth and escape.

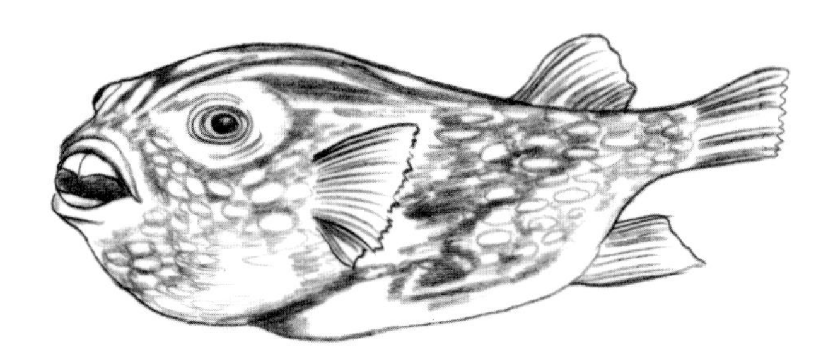

Some fishes feed on clams, crabs, corals, and other hard-shelled creatures. They are likely to have broad, platelike teeth. The puffer is one of these fishes. It can crack the shells of clams with its teeth.

The barracuda lies in water without moving until it is ready to strike. Then it seizes its prey with its long canine teeth. Its smaller, daggerlike teeth cut the prey to ribbons.

The barracuda is one of the fishes that has another fish as a cleaner. The cleaner is a wrasse, a small, brightly colored reef fish. It feeds by cleaning the barracuda's mouth. It eats the bacteria that live there. The wrasse gets a meal. The barracuda's mouth is kept free of disease. A wrasse almost never ends up in a barracuda's stomach.

The jaws of a tiger shark are edged with jagged cutting teeth. When the shark attacks big prey, its powerful jaws close on the victim. Its teeth slice into the flesh. The shark shakes its prey and a chunk of flesh is cut away.

A tiger shark's teeth are not rooted in the jawbone the way yours are. They grow in its skin. Sometimes teeth are torn out of the mouth. They also wear out or fall out with age. Then other teeth take their place.

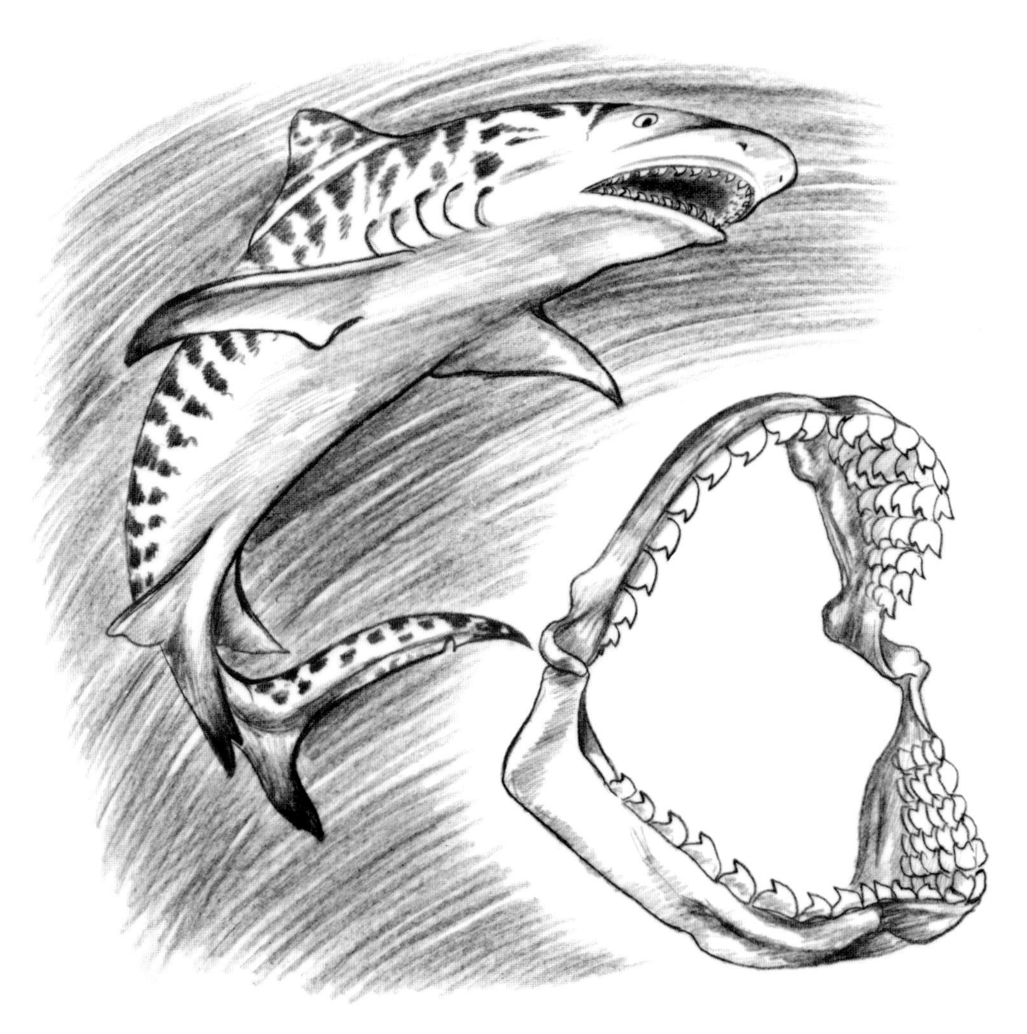

A tiger shark has several rows of spare teeth. They grow behind the teeth that are in use. The spare teeth are flattened inward and overlap like the shingles on a roof. When outside teeth are lost, teeth in the next row spring up and move into place. A new row of teeth starts to form behind the last row. In five years a tiger shark may form, use, and lose 12,000 teeth.

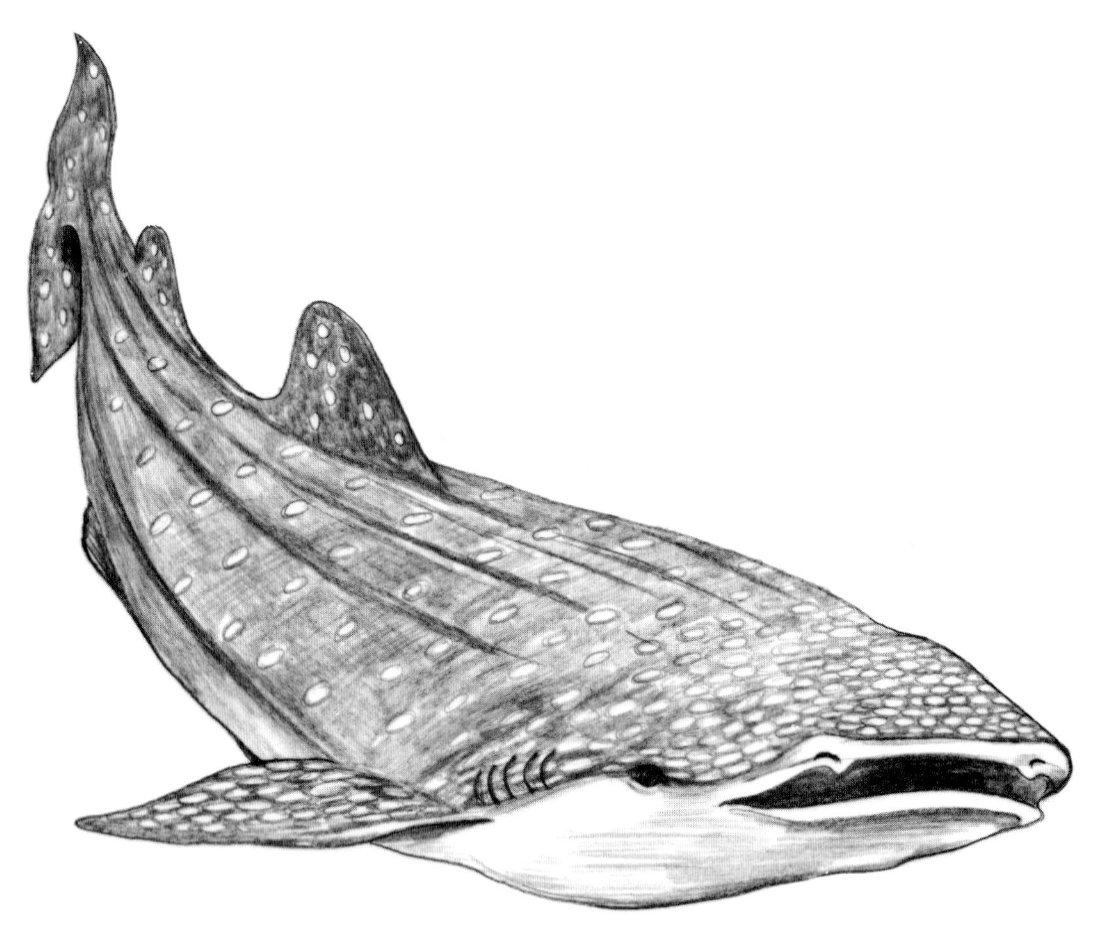

Many kinds of fishes have small, weak teeth or no teeth at all. They feed by straining food out of the water. The biggest fish of all, the whale shark, feeds in that way. It has 7,200 teeth, but each is only an eighth of an inch high. The teeth may trap food that flows into the mouth, but they are not useful for catching it. A whale shark swims along with its mouth open, taking in water and food. The water flows out over its gills. The food is caught by strainers called gill rakers.

In the Days of Dinosaurs

Teeth are clues to what animals eat and how they get their food. Teeth are clues to an animal's relatives. They are also clues to the past—to animals of long ago.

When an animal dies, the soft parts of its body usually rot away. The hard parts—the bones and teeth—are left. In time they may wear away or crumble into dust. But this does not always happen. Sometimes they are covered by layers of sand, mud, or other material. Over many, many years, more layers are added. The top layers press down on the bottom layers. Slowly the bottom layers turn into rock. The bones and teeth are sealed in rock and preserved. That is, they become fossils.

Fossils tell us that once dinosaurs roamed the earth. Most kinds were plant eaters, but some ate animals. How did scientists learn that? You know the answer—they could tell by the teeth. Teeth are also one of the main clues that scientists use to sort dinosaurs into groups of relatives.

This dinosaur had teeth that were six inches long and had edges like saws. They were set in four-foot-long jaws that opened wide. The teeth belonged to the largest meat eater ever to walk the earth, *Tyrannosaurus rex*.

Diplodocus was a giant plant eater, 85 feet long. It had peg-shaped teeth, good for nipping off buds, fruits, twigs, and leaves. Its long neck let it browse among the treetops. Diplodocus's teeth show signs of heavy wear—and no wonder. To stay alive this huge plant eater must have spent most of its time feeding.

Diplodocus had only front teeth. It had no grinding teeth, and that is a puzzle. How did its food get broken down? No one knows for sure. But one idea is that *Diplodocus* may have handled food the way birds do today.

Birds have no teeth. Most of them swallow their food much as they find it. But a bird has a special stomach that is called a gizzard. The gizzard has walls of strong muscles and a lining that is hard and ridged. The gizzard takes the place of teeth. It is a kind of mill, where food is ground up with the help of the small stones that most birds swallow.

Did *Diplodocus* and its relatives have gizzards? One clue shows that they may have. Large numbers of polished stones have been found with some skeletons. It is possible that these stones were in gizzards. As they helped to grind up food, they became polished and smooth.

The duck-billed dinosaurs were also plant eaters. They had hundreds of teeth. The teeth in each jaw were pressed together, forming a rough plate. Food was ground up between the two plates.

The food must have been hard and gritty, for the teeth show signs of heavy wear. They were always being worn down and spat out. New teeth grew in to take the place of lost ones.

In the last days of the dinosaurs, giants were eating giants. How could anyone know that? One way is by the tooth marks found on bones. Here is a brontosaur whose tailbones were gnawed by an allosaur—the tooth marks fit with an allosaur jaw and teeth. While feasting, the allosaur lost several teeth. They were found near the skeleton. Other skeletons have the broken-off teeth of meat eaters stuck in the bones.

Look at these animals and their teeth. You may still think, "What big teeth you have!" But now you also know that teeth tell many tales about animals and the food they eat.

Index